O B J E C T
TELEPORTATION > T >

OBJECT
TELEPORTATION > T >

PROVEN DISCRETE & AUTHENTIC

VIJAY JAIN
DISCOVERER & INVENTOR

PARTRIDGE
A Penguin Random House Company

To order additional copies of this book, contact
Partridge India
000 800 10062 62
orders.india@partridgepublishing.com

www.partridgepublishing.com/india

Mother of Vijay Jain

Father of Vijay Jain, Mr. M L Lunia

Vijay Jain, the author

Diana, Mrs.Madhumita Bothra Sacheti, my
beloved niece and Science inspiration

NO…. IT IS NOT A DREAM TALK ANYMORE SINCE PEER REVIEW JOURNAL NATURAL.

CONFIRMS IT BY TWO HISTORICAL LETTERS TO ME FOR TWO DIFFERENT BLUEPRINTS FOR OBJECT.

TELEPORTATION SUBMITTED FOR EXPERIMENTS.

> ROCKET, JETS,…….ALL SUCH WILL NOT HOLD MUCH – JUST AFTER

  "TELEPORTATION ADVANCE ONE" – GIVE NEW

  DIMENSION TO MANKIND.

➢ SERIES OF SYSTEMS OF DIFFERENT WILL BECOME ABSOLUTELY OBSELATE GIVING UNNATURAL WORLD BUT IN A NATURAL WAY.

Honorable and Respected

Vijay Jain
VI Nundy Street
Kolkata 700029
vijay_jain27@yahoo.co.in
97484 99180/98317 47530
033 – 6458 2200

It is a privilege that I have discovered the method of Teleportation and it published in the Newspaper. The Telegraph of Kolkata in how column on 8 may 2000. Again I became successful in sending the blueprints for the invention of two types of teleportation and it has been highly acclaimed international scientific community alike.

I am working with best of my ability for Experimental Mouse Teleportation. I should have been happy for such great achievement up to a sky height, But alas, by negative use of Teleportation and surely` Instaportation instead of boon –Teleportation can be a last game on the planet earth. Very very shamefully, I declare that I have theory in my mind by which our universe can end in

Zero second (The zero Second Infinity Disaster Theory)

International heads must assure and insure we Scientist that NEVER NEVER – EVER. There will be a single case of negative use of Teleportation and I sincerely hope that this condition of we Scientist won't be Turned down.

<u>Thanking You,</u>
<u>Sincerely</u>
<u>Vijay Jain</u>

T>
OBJECT
TELEPORTATION
DREAMY OR
ABSURED
???
  ➢ JUST
    ACCESS!
  ➢ VIJAY JAIN

<u>NPG nature publishing group</u>
<u>The Macmillan Building</u>
<u>4 Crinan Street</u>
<u>T +44 (0)20 7833 4000</u>
<u>F +44 (0)20 7843 4640</u>
<u>www.nature.com</u>

<u>OBJECT</u>
<u>TELEPORTATI</u>

Dear Author

Thank you for submitting your manuscript, which we are regret fully unable to offer to publish.

It is Nature's policy to return a substantial proportion of manuscripts without sending them to refers, so that they may be sent elsewhere without delay. Decisions of this kind are made by the editorial staff when it appears that papers are unlikely to succeed in the competition for limited space. In the present case, while your findings may well prove stimulating to others thinking about.

Such question, I regret that we are unable to0 conclude that the work provides the sort of firm advance in general understanding that would warrant publication in Nature. We therefore feel that the paper would find a more suitable outlet in specialist journal. I am sorry that we cannot respond more positively on this occasion, but I hope that you will rapidly receive a more favorable response elsewhere.

Nature Administration

Honorable & Respected,

I have already sent 14 Research papers in hard copies and few in webs too. This particular paper which I am submitting is being considered very very important by me.

I will be grateful to those who can make this blueprint for experimental in an immaculate and precised manner with a touch. Some rectification which I am sure would be needed without any doubt.

As usual, my papers are never have old styles or old methods or decorated diagrams and figure sets etc hence please bear with me and forgive.

I am confidential but at the same time I am also sure that without the assistance of typical Experiment like these or the other – Teleportation is not going to be a success for along time to come.

I request Scientist and liberal minded to contribute to make the Teleportation a success by getting inspiration from persons like me or any in the world but to catch up the target with very high imagination and independent creativity of one's own and not mere ECHO of some kind.

<u>Thanking You</u>
<u>Yours Sincerely,</u>
<u>Vijay Jain</u>
<u>VI Nundy Street</u>
<u>Adjt building: Railway Ticket</u>
<u>Booking Counter</u>
<u>Gariahat Kolkata -700029 India</u>
<u>Contact-97484 99180</u>
<u>Email-vijay_jain27@yahoo.co.in</u>

PART I

THE TELEPORTATION
BLUEPRINT

Beam up, Scotty'

Matter is made of atoms and atoms are made of protons, electrons and neutrons. If an object's atomic blueprint can be read, transmitted ales where and duplicated by some means. If amounts to teleportation of the subject. We have already made some advances in that direction, with the invention of the telegraph, telex, etc. Traversing the rest of the way should only be a matter of time.

Vijay Jain
Calcutta

8 MAY 2000
The Telegraph
Know How
Third aw
Calcutta
India

TELEPORTHETATION BLUEPRINT

Teleportation Specific:

If an object's Atomic Blueprint can be read, transmitted elsewhere and duplicated by some means, it amounts to teleportation of the object.

Atomic Blueprint of a BAR MAGNET

A small bar of torn is being chosen for getting its atomic blueprint for Teleportation. An application of photoelectric effect on bar of iron is to be applied. There will be immediate release of Electron would take place due to electromagnetic radiation of sophisticated type & system would be applied.

A very bright light having high frequency would cause high emission of electrons from the iron bar. Since the electron released – the total electrons – can be stored in a separate chamber which will be containing the electrons only in vacuum.

Now we count the total electrons the remaining protons and neutrons can get accessed too. The atomic no of iron is 26. Now if we can have the total particle which an iron bar is having then after the 100% assessment of individual particle – Electron Proton Neutron we are to choose heatproof glass replica of iron bar and punch the raw Electron proton neutron in an accurate quantity of each as per the assessment done.

Iron bar is being subjected to enhance back the electrons arrests and be a normal iron as before.

Iron Bar – ALPHA

REPLICA GLASS CONTAINING PARTICLES–BETA

Alpha and Beta is to be kept very near to each other and should be heated up simultaneously in the uniform temperature at a very higher side. At Alpha – Ferromagnetism would take place.

Ferromagnetism – Q is for quantum – by john gribbin page 133.

The kind of magnetism associated with an bar magnet. A bar of ferromagnetic material can be thought of as made of enormous number of tiny eternal magnets (corresponding to group of Individual atoms). When the ferromagnetic material is hot, these tiny eternal magnets spin around the jostle one another at random, so there is no overall magnetic field, in this state, the material is said to be paramagnetic. But when the bar cools at a certain temperature (known as curie temperature or curie point) the internal magnets curie suddenly line up with one another so that there is overall magnetism with each of their tiny magnetic fields adding up to produced the familiar overall magnetism of a bar magnet.----------------------.

Under DOMAIN a considerably strong magnetism is going to take place at – Alpha

BUTINFLUENCING – BETA at the same time now Beta's position is like copycat that Alpha in a systematic and calculated as well as sequence wise manner. After paramagnetic stage when we will allow the iron bar about to be called – A magnetic Bar – when it comes down from paramagnetic stage to curie point.

The reaction at Alpha will influence the Beta due to attraction due to induction as soon as the magnetic field of alpha chases the nearest BETA hence the phenomenon passes through hysteresis cycle giving the effect to Beta too but Beta itself is having the effects within.

Under a nascent stage of Beta when it is just and just going to settle down as – about to magnet to be – in an atomic magnet stage it should be transmitted by most modern in a speed of light at the desired place.

I want to name this transmission as

DIANA TRANSMISSION

Duplication at the other end is not involved because the atomic staged bar magnet – when it will get below curie point, automatically – A physical bar magnet is bound to be there surely done by

- TELEPORTATION

NPG nature publishing group, The Macmillan Building
4 Crinan Street
T +44 (0)20 7833 4000
F +44 (0)20 7843 4640
www.nature.com

Dear Dr. Jain

Thank you for submitting your manuscript, which we are regretfully unable to offer to publish. It is Nature's policy to return a substantial proportion of manuscripts without sending them to refers, so that they may be sent elsewhere without delay. Decisions of this kind are made by the editorial staff when it appears that papers are unlikely to succeed in the competition for limited space.

In the present case, while your findings may well prove stimulating to others thinking about. Such question, I regret that we are unable too conclude that the work provides the sort of firm advance in general understanding that would warrant publication in Nature. We therefore feel that the paper would find a more suitable outlet in specialist journal.

I am sorry that we cannot respond more positively on this occasion, but I hope that you will rapidly receive a more favorable response elsewhere.

Nature Administration

Cell:
91 33 97484 99180//98317 47530
Vijay Jain
VI Nundy Street

Kolkata 700 029 India
Vijay_jain27@yahoo.co.in

Honorable and Respected,

I am sending my latest papers on
1. Experimental Blueprint for any metallic object Teleportation.
2. The Teleportation Blueprint – previously sent to support the latest on teleportation.
3. Conversion due at neither mass nor energy point.

Please note that if any flaw which is bound to be, is being detected them get it right and if it gets published them please mention the name of that particular editor who does so. I hope that you will consider this particular suggestion of mine in crucial cases especially. Well I have much that a high tag scientist to get fame and throw my discovery and invention via blueprint, in Arabian Sea. I wish that a Scientist should get a pat in the back whenever he does or tries to do something great in true sense and hence I am obliged for life long who did not print that is published my paper on object Teleportation but i stand – as Mr. Right according to letter given to me by nature.

CONVERSION DUE AT NEITHER MASS NOR ENERGY POINT

As we know that mass is the matter it contains and it is a fixed quantity. We also know that mass and energy is mutually convert able (EINSTEIN) hence present matter deals with the specific Mass

Energy getting converted?

No, but

In between,

Mass has neither changed into energy nor energy has harnessed the mass to change that is mass is mass and about to change into energy but it has sucked up between Mass/Energy junction the pin point of that a mass is not mass in true sense nor energy is not borne in true sense. 'c' the universal constant is yet to be borne though in our experiment we have already excited the mass to exchange its form but our experiment lies in the mystery of it when it is in – in between position.

1. It is somewhat – volatile whenever mass just changes into energy.

2. Origin of 'c' can get challenged if some desired or calculated amount of ray, compressed particle, nutrine, etc is subjected to pass through when mass is just and just changing into energy – that when 'c' is not borne but about to and we do the experiment at.

3. No reaction that is in this case also – the conversion under normal circumstances cannot take place without the presence of gravitational field how feeble it is and we should keep in mind that the machine or a mass of any type by which energy is being evoked also has certain degree of mass – there by gravitation and its field.

4. 7 colors which has different intensities can be subjected alone or with different type of mixtures as particles to mass – energy pin point.

5. All mass represents energy (EINSTEIN) mass wants to change into energy and vice versa. This particular can prove deadly beside many if an

experiment can be carried out with various angle and definitely – we are bound to get strange results not known yet.

6. Lastly, one student of mine declared : It can be birth of UNIVERSE. I was thrilled up fully, but could nor reach to any conclusion. (The name of the student – I must mention is Mr. Viswanath Das who is Government Employee).

Experimental Blueprint for Any Metallic Object Teleportation

After successful blueprint of iron bar teleportation (paper attached) I am submitting another paper for any metallic body teleportation for your consideration. Step by step, we are getting closure towards all type of matter teleportation but in any type of technique which is swift and without hassles and a cumber son process.

In the given figure, A and B – The Strong MAGNETIC PILLAR which looks like bracket right from mathematics but they are actually electro magnets of such a nature that intensity of ' B ' is greater than that of ' A ' s far as magnetic field is concerned.

In our case, we are selecting the

GOLD BAR

Whose replica can be teleported at a desired place. By applying Geigerseillitation or Thallium activated sodium Iodide – we can access the complete atomic and particle potential of that particular Gold Bar which is being used in our experiment.

A heat proof non breakable glass is being taken and the particle assessed from gold bar at ' A 'the same amount of it (Electron, Proton, Neutron) is being punched in ' B '.

Now A & B both are being heated up by measured electric at a high temperature.

The Gold Bar will get melted at ' A '. In ' B ' – The particle Electron, proton, neutron will badly Collapse with the walls of the tube in a volatile manner.

Now the heating to be stopped at ' A ' and ' B ' but we have to keep in mind that ' A ' is having less Intensity than ' B ' . Hence as gold bar begins to settle down after heating part is over but in a new position though now Electro magnets are in on position.

Since magnetic field is in force now hence the particle at 'B' (Electron at 'A') will(repeat electron At 'B') hence, as the particles will begin to settle down in both the glass tube similar in position to Each other – keeping in mind that intensity if magnetic field at ' B 'is greater than 'A'.

Wall P and wall Q should have removed for sensitive purpose for the attraction to each other that Is between 'A' and 'B'.

At a balanced and equilibrium stage as usual a transmission – DIANA TRANSMISSION – should be done at a desired destination and we will get a GOLD BAR after Curie point is being attended and the Teleportation is or can be said as complete.

❖ AFTER HEATING PART IS OVER, CURRENT AT ELCTRO- MAGNETS, SHOULD GET "ON".

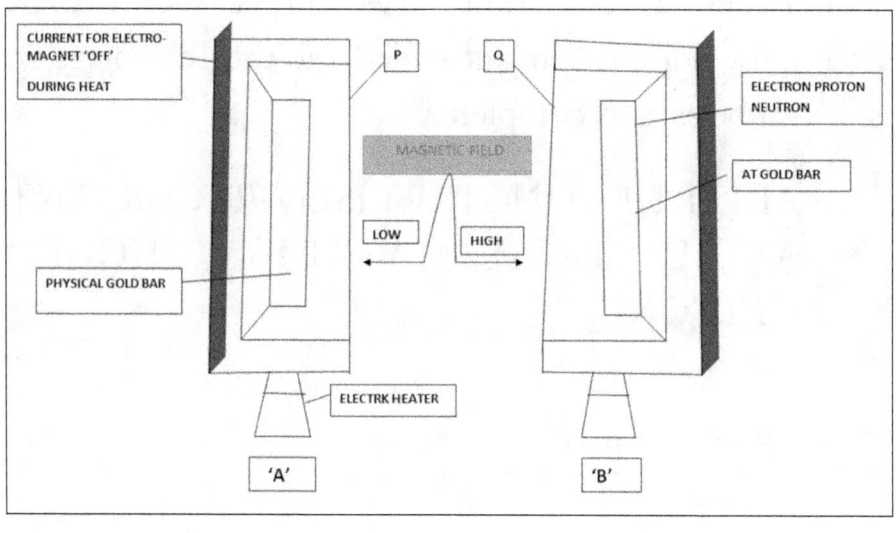

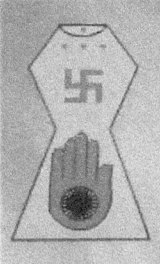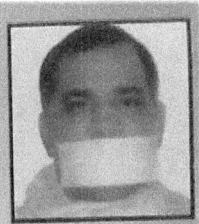

Muni Shree Mani Kumarji

(Astrologer, Palmist, Vastu & Avadhankar)

SHREE MAHENDRA MUNI MISSION TRUST

4C, Town Shend Road (Bhawanipur) Kolkata-700 025, India
Phone : 91 33 2476-3537, 2476-4691
(Near Ramesh Mitra Road Petrol Pump)
E-mail : munishreemanikumarji@hotmail.com
E-mail : munishreemanikumarji@yahoo.com